Solar PV Water Pumping:

How to Build Solar PV Powered Water Pumping Systems for Deep Wells, Ponds, Creeks, Lakes, and Streams

by Christopher Kinkaid

 Solardyne.com

Published by Solardyne, LLC
Portland, Oregon

ISBN-13: 978-1500445232
ISBN-10: 1500445231

Table of Contents

Preface

Water pumping is a big job. Solar electric (PV) powered water pumps, are the most effective way to pump your Deep Well, or shallow pond, river, lake or stream. Solar water pumping systems, pump water, with high performance, reliability, and no fuel-costs. Is your Well, Pond, or Lake at a remote site? Solar electric photovoltaic (PV) panels, at historic low prices, lower costs, and can be your water pumping solution.

Water your livestock, irrigate your orchards, gardens, fields, or farmlands with this Easy Step-by-Step Guide complete with specific examples of water pumping equipment for different situations. Pump water from your well, or shallow surface source directly with Solar PV panels. Size your solar water pumping system with this Step-by-Step Guide to defining, and building your solar water pumping project.

Solar water pumping systems include solar PV panels, a controller, and the Pump itself. The examples, included in this book, match the model pump, with the solar power supply required to

produce a given average water
production per day.

About the Book

This Book is written as a step-by-step guide to defining your solar water pumping project's "vital statistics," and choosing the right equipment to get the job done. If you have a specific solar powered water pumping project in mind, then visit the Solar PV Powered System Examples List located at the Quick Guide in Chapter Nine.

The **Quick Guide** takes you different, specific, Solar Water Pumping Systems. The Solar Water Pumping examples are defined by Depth of Well, and Gallons Per Day delivered. If you're pumping from a shallow source of water, such as a pond, creek, lake, stream, or small river, the systems are Listed by Gallons Per Day delivered.

In **Chapter 2** outlines the Step-By-Step process to define your system for your own system design, or to speak with an outside vendor. Use this process to determine the "vital statistics" of your system.

Chapter 3 discusses the use of Solar Power Supplies, and how the listed examples are configured in this eBook.

Chapter 4 through 7 describe water Well pumping using Submersible pumps which range in Depth from 20 feet to 800 feet. System examples, include Solar PV Power supply parts lists describing the specific solar PV panels you'll use, and at what System Voltage to operate your pump for highest productivity.

Chapter 8 outlines pumping water with Solar Power for Shallow Sources of water such as ponds, lakes, streams, creeks, and small rivers. Solar PV Systems are defined by total "Rise" or lift, such as from hills and small escarpments on your property, and the total "Run," or Distance horizontally you want to move your water. Solar systems listed can pump as far as 4 miles, and lift as high as 400 feet.

This book "Solar PV Water Pumping" was written as a resource for planning and implementing a Solar electric (PV) powered water pumping system, to deliver water for remote sites. Ideal for remote cabins, remote homes, off-grid living, garden, orchard, agricultural and livestock watering projects, solar PV panels make an excellent power supply and can pump great amounts of water.

Introduction

The need for pumping water is fundamental to life, and predates the Neolithic age. Without moving water, there is no civilization. Then, as now, our demand for water is vital for agriculture, livestock, residential, commercial and industrial needs, and being available at your site everyday, solar energy can be an effective power supply to great advantage.

Today, modern solar electric panels (PV) make pumping water relatively easy to install, cost-effective, and offers outstanding performance and reliability where it counts: day-by-day in the field. Solar PV panels are solid-state, have no moving parts, hermetically sealed from the environment, ruggedly framed, rated for extreme locations and often carry 25 year warranties, making for a reliable power supply.

With proper design, and hardware choices, (the point of this Book), solar water pumping systems are surprisingly productive lifting water from great depths, and/or moving water great distances with respectable flow-rates.

This Book is intended as a Step-by-Step guide to first define your solar water pumping system, then match that project to one of the examples provided. If you need more water pumped than the sample systems listed, use **Chapter Two** to define your project so that your Water Pump supplier can quickly identify the right system for your specific project. Solar PV panels provide strong DC voltages which match very well to the DC solar pumps available in the market.

Use Solar PV power to pump your Water Wells from 20 feet to 800 feet. Use Solar PV Power to pump water from your pond, lake, creek, stream, or small river using Surface Pumps. Have a Solar Water Pumping project in mind? Visit Chapter Nine for a **Quick Guide** to the included systems.

Chapter One - Solar Water Pumping the Big Picture

Solar powered water pumping systems can lift from deep sources of water, like deep wells directly, as well as, pump from shallow sources like ponds, lakes, creeks, streams, and small rivers. There are two basic-types of solar powered water pumping systems, depending on your source of water: Wells, or Surface Sources.

In this book, we'll break down the questions you'll need to ask to define your system requirements. Then, we'll match those requirements to the appropriate Solar Water Pump type and specification to get the job done.

Need to lift water over 600 feet? Need at least 7,000 gallons delivered to a cistern 200 feet away from the pump? This book will step through whats needed to work out these aspects and arrive at the best solar water pumping system for your particular water pumping project.

We begin by defining your daily water demand - how much water do you need each Day? What is your water source depth? We need to know some basic information about your water type beginning with the Source of your Water. Solar water pumps use different equipment depending on the source of water whether from well or surface water.

For deep well pumping the standard pump type used is the Submersible pump. The Submersible pump needs a well of at least 3" inches in diameter, (4 inches for larger pumps) and is dropped down into the well with the power cable, drop rope, and water delivery hose. The Solar electric PV panels, mounting racks and the controller is mounted above ground near the well head, only the submersible pump and cabling/tubing is dropped down the well.

Surface sources of water, typically shallow, such as lakes, ponds, streams, rivers, cistern, or tanks will use a Surface Pump. Several types of surface pumps exist depending on how much water you wish to pump, each with advantages and features.

Later in this Book under Surface Pumps we'll go through the different features of each type, and how to analyze the "Water Quality" of your shallow source. Ponds, lakes and other open systems can be cloudy, or turbid with particulates in the water making it gritty. Some surface pumps are vulnerable to gritty water. If you're water is turbid or gritty then an In-Line filter will be required.

Surface solar powered pumping systems mount the Solar PV panels, racking, controller and the pump itself all top-side and mounted a few feet from the water source. Surface pumps are placed next to a creek, pond, or stream where its usually shady due to trees or bushes. In this case, the solar PV panels can be placed up to 75 feet away from the pump.

The surface pump needs to be placed onshore, near the water (within 10 feet horizontally, and 10 feet vertically), and on a firm foundation.

Laying a small cement pad is not a bad idea if you're going to leave the pump for long periods. If you're in an extreme climate, then you should build a box, or have an outdoor enclosure to protect the pump, and the pump controller from the elements.

Once your surface pump is installed near the water, only the intake hosing is submerged into the water source. Surface pumps are different from submersible pumps as we'll see later in the book.

Rise, Run, and Water per Day

All water pumping projects can be defined by three basic water factors: **rise**, **run**, and desired **water volume** delivered each day. Once we have these aspects defined we'll work the load backwards and arrive at the properly sized solar power supply to run the pump. The "Rise" refers to the total height (head) that you need to lift the water. Your source of water could be a well, for example, you know the water table is at 100 foot depth. You may also need to lift that water an additional height to fill your tank or cistern. Add all of these heights to reach your total "Rise."

"Run" refers to the length of distance you need to pump your water on the surface. Even though your land may go up and down, the Run refers to the total length of horizontal distance you'll need to pump to reach your tank or cistern. Next, you need to have a number for the total Daily amount of water you need to deliver.

Many pumps are rated by the Gallons Per Minute (GPM) they pump. This can be a mis-leading value, as unlike a plug-in AC pump which can run as long as you like, there is a limit to how many hours per day your solar PV panel will power the pump. Therefore, think in terms of how many Gallons Per Day (GPD) you'll need not just in terms of Flow-rates but total amounts for each day's production.

For example, the demands of watering cattle can be estimated at 30 Gallons per Day, per Head of cattle (more if in a hot climate). A herd of 200 cows requires 6,000 Gallons per Day. Be sure to estimate your water needs in terms of Gallons Per Day (GPD), this will help you size the solar powered water pumping system you need for your application.

Solar energy is a powerful force. The intensity of the sun in any given hour will fluctuate being a natural source, and for water pumping this is important, but over the course of time the sun delivers a reliable average of energy. The solar peak power (1,000 watts per Square Meter) is used to estimate the actual energy delivered by a solar PV panel for the purposes of pumping water. Every location on Earth has an equivalent Solar Peak-Hour equivalent. In Portland, Oregon the Peak-Hour rating is 3.5 Hours per day. In Kansas, the solar Peak-Hour rating is 5.5 hours, for example.

For your projects location do an Internet search for your sites peak-hour rating. Multiplying the Solar PV panels power rating by the Peak-Hour rating of your location tells you how much Energy your solar PV panels will produce, on average, each day at your site.

Example 1: If your pumping site is in Kansas, with a 5.5 Peak-Hour rating, then 1,000 watts of Solar PV power will produce 5.5 Kilowatt-hours (kWh) of energy per Day.

Example 2: If your pumping site is in Southern California (6.5 solar Peak-hours) using a solar PV panel power rated at 500 watts, how much energy will that 500 watt solar panel produce? Answer: Energy equals Power x Time. The power panel rating (500 watts) times the Peak-hour rating (6.5 in this example) produces a daily energy production of 3,250 watt-hours, which equals 3.25 Kilowatt-hours (kWh) each Day.

Solar Water Pumping in the Field

Solar powered pumps can operate in diverse locations including deserts, tropics, high-altitude, cloudy and urban environments. If you're sizing your own solar power system the Power-Rating of a solar PV panel must be "De-rated" depending on these extreme conditions. For example, all electronic devices dislike heat, higher temperatures cause a voltage drop in PV modules. Solar PV panels, by definition, are in the sun and can become very hot.

If you're in a particularly hot location derate your Power-Rating by 20%. In the examples given in the Chapters below the necessary De-Rating has been calculated so if you follow my examples you're all set. If you design your own systems then be sure to derate the solar panels.

Once you know your Rise, and your Run, the next key is to know How Much Water you need for Each

Day. Once you know your Daily Need expressed in Gallons per Day (GPD), then we can begin to work the problem backwards ending up with the right equipment for pumping your water.

If your source of water is in a remote location, and electricity is either unavailable, or expensive to wire, solar power makes an effective choice. Grid powered water pumps use Alternating Current (AC). Solar powered water pumping systems, in contrast, use Direct Current (DC), giving an excellent match to solar PV panel, and battery voltages.

Traditional AC pumps that run off of traditional Grid power are most often "centrifugal" pumps and are designed to spin at very high speeds pumping as much water per Minute as possible. Typical AC pumps have high power consumption power draws, especially when facing high pressures (often self-induced by pumping more than the pipe can manage), or in the case of very low flow-rates, resulting in lower efficiency.

These issues make Solar powered water pumps an attractive choice from a performance perspective, as DC voltages from your solar array are designed to closely match the draw of the pump. Further, controllers act as Maximum Power Point Trackers (MPPT), which further increase the efficiency of DC solar water pumping.

To maximize the performance of DC power systems solar PV powered pumps are often built of more efficient pumps, and use "positive displacement-type" technology which pumps a set amount of water with each rotation of the pump blade. Cloudy and inclement weather may present less power from the sun at any given moment, but the positive displacement pump won't suffer any efficiency loss at low power. Therefore, if you only have half the sunlight, you'll still pump half the volume of water. Excellent range of efficiency for real world conditions of changing light levels.

AC pumps are designed to push as hard as possible with the aim of pumping more water as fast as possible. However, these high power electricity hungry AC pumps produce a high amount of "internal" friction inside the pipe wasting energy. The smaller the diameter of the pipe you choose, the more internal friction will exist for a given water speed. Slow pumps, as you'll see later in the Surface Pump chapters, take great advantage of moving your water slowly through the pipe greatly increasing efficiency. This minimizes internal friction, and lowers the size of the solar PV array necessary to power the pumping system.

The solar water pumping DC strategy verses an energy hungry AC plug in pump, is the classic race between the tortoise and the hare. The AC pump is the hare, pumping a large amount of water in a short time. The solar water pumping DC system is designed to be the tortoise, and over the course of the day, deliver the amount of water you expect from the system. This advantage translates into great savings in the cost of your system by making it smaller.

Submersible Pumps for Pumping Well Water

If your water source is a deep well, then you'll need a Submersible pump. Well water pumping with a submersible pump, powered with solar PV panels can deliver from 1 Gallon per Minute (GPM) to over 80 GPM using direct solar power. The larger the Solar PV panel array, the more water you'll pump.

The amount of water you can pump with a given solar PV panel array will depend on the total Rise, (Height, Head) you'll need to lift the water. Be sure to analyze your water table within your well and estimate if your water table drops as you pump out water. Most wells will drop in water table a little, or more under some conditions, during pumping so you'll want to estimate your well depth with a margin of error to compensate. This is the depth you will lower your submersible pump to with a drop-line, (usually rope, or cable).

Submersible water pumps are designed for the harsh conditions of being underground. The cooler temperatures of the water at these depths help keep the pump running cool and extend the life of the pump.

If you're going to use a submersible pump to pump water short vertical heights from ground cisterns or ground tanks to a roof tank, for example, then some protection from over heating the pump must be used. If you're going to pump from a ground tank, to the roof (only 25-35 feet vertically), and you wish to use a submersible pump, then mount the pump within (concentrically) a large vertical plastic Pipe which acts like a chimney.

The Pipe is of a larger diameter than the pump to allow water to flow up and around the pump. The "height" of the plastic pipe will be slightly longer than your pump, with your pump in the middle. The idea is that water taking heat from the pump will have a direction to go, up, bringing in more water from the bottom of the Pipe, cooling the pump. Submersible pumps in deep wells have no problem overheating and are designed for their operating conditions.

This book will cover different well depths and water amounts with the proper solar PV power supply parts list in the specific Chapters below. You'll choose your solar powered Submersible Pump based on the depth of your well (Rise), you're running distance (Run), and the total amount of Water Per Day (GPD) that you want to deliver.

Solar powered submersible water pumps can be designed for smaller systems and can be powered with as little as 200 watts of solar PV. Submersible water pumps such as the SHURFlo 9300 and Aquatec SWP-4000 are built to be powered direct by PV solar arrays from 100 to 200 watts, respectively.

These SHURFlo and Aquatec models of Submersible pumps can deliver from 500 to 1,000 Gallons Per Day (GPD) lifting water up to 200 feet.

Deeper wells up to 800 feet are best served with Submersible water pumps such as the Grundfos line and are rated for higher lift capacity, higher water flow rates, and does not, typically, require service for 15 to 20 years, with proper installation. Grundfos makes the SQFlex line of Submersible pumps. If you're going to pump from a well up to 800 feet deep, and need larger amounts of water, use a Grundos Submersible pump. The no maintenance, long-lived pump will save you in field maintenance, time, and effort pulling your pump.

Solar Pump Controllers

Nearly all solar powered water pumps need a pump controller wired between the Solar PV panel and the Submersible pump. **Controllers** sample the voltage and current produced by your solar PV panel, and match it to the actual load on the pump. This increases efficiency dramatically. The controller is the "**brain**" of your system and ranges from a simple on/off switch, to a an intelligent system that monitors your operation and alerts you to over-current, or run-dry conditions and will stop your pump.

Larger solar powered submersible pumping systems such as the Grundfos SQFlex submersible pumps can be powered directly from your Solar PV panels, or small wind generator (48-300 VDC) through the right controller.

You can also power your SQFlex submersible pumps with an inverter, generator, battery, utility grid, or any combination of these power sources, as a back-up power supply. The SQFlex line of submersible pumps can be run by almost any source of DC power from 30 to 300 VDC, and 90 to 240 VAC using Alternating Current sources. To do this the Submersible pumps require a "Controller" to manage the power to the pump.

Using just solar PV panels, the SQFlex submersible can be controlled with the IO50 control box. This controller features a simple manual on/off switch which mounts between the solar PV panel and the submersible pump. This allows you to turn off the DC power from the Solar PV panel reaching the submersible pump when you're installing, inspecting, or servicing your pump.

For greater control of your submersible pumping system use the **CU200** Interface box. This controller allows you to communicate with the pump, and monitor different aspects of your pumping system. If you wish to add wind, battery, generator, AC grid, or other power options you'll need the CU200 interface.

There are many advantages to the CU200 including built-in diagnostics to give you operating status, power consumption, and allows you to connect a Water Level Switch. The remote Water Level Switch is a float-switch that turns off your pump when your tank is full. (Some pump controllers allow you to have several float-switches to also start up your pump when tank levels are low).

Controlling your solar water pump with a float switch is a great option. The float-switch can be mounted in your tank, and can be placed over 1,600 feet from the pump controller. Note: (Use 18 AWG two-conductor wire if you're running your float switch this far from the controller).

If you're going to hook up a back-up generator to power your pump in addition to solar PV panels used for normal use, you'll need the IO101 AC Interface box. You can use a generator as a back-up, or you can use the AC grid, if available, as a back-up source of power. This interface box control is limited to 120 VAC outputs so only single phase AC inputs can be handled. Back-up diesel, or gas powered generators are usually sized between 1.5 and 3.5 Kw for running these SQFlex submersible pumps.

Solar powered submersible pumps like strong voltage. Voltage is the electric "pressure" produced by the solar PV panels. The minimum voltage you'll need from your solar PV array is defined by what your Pump Voltage needs, and is usually 12, 24, 48, or 96 VDC.

The minimum voltage for 48 VDC pump, most common for deeper wells and surface pumps is 30 VDC under load, but wiring for 100 VDC is most efficient for the maximum performance from your pump.

Solar PV panels can be wired up to 600 VDC in series, but solar water pumping systems operate best around 100 VDC, therefore wire your solar PV panels in series to 96 VDC, ideal for deep wells. Solar PV panels come in many sizes and Power ratings. Smaller Solar PV panels from 5 watts - 80 watts are usually hard-wired as 12 VDC modules.

For powering a smaller submersible pump using smaller PV panels you'll wire your panels in "Series" to increase voltage. Two 12 VDC panels wired in series produces 24 VDC. Wire four 12 VDC solar PV panels in series for 48 VDC. This is a good operating voltage for small pumping systems.

Surface Pumps for Tanks, Cisterns, Ponds, Lakes, Streams, and Small Rivers

Shallow sources of water such as ponds, streams, lakes and small rivers can be pumped with solar PV power very well, but have different demands than submersible pumps. For pumping shallow sources of water you'll use a Surface Pump. Surface pumps have several types, but in all cases are mounted near the source of water, slightly above the water, and on a firm foundation.

Many orchards, gardens, and fields, for example, are watered from a storage cistern, or tank positioned above the field so water can be gravity fed to the plants by opening a valve. Pumping water from a nearby creek, running at a lower elevation than the cistern, presents a typical water pumping scenario. A solar powered surface pump would be used to push the water from the source creek up to the cistern. In the Chapters below examples are included for different Surface Pumping solar powered systems and scenarios.

Surface pumps can push water uphill and through long distances of pipes to fill cisterns and storage tanks, and to pressurize water tanks for irrigation and livestock watering. Be sure to place your Surface Pump no higher than 10-20 feet above the source of water, and closer is better. Pumps are designed to push, not pull. Since atmospheric pressure is about 15 psi the vacuum a pump can draw is limited to this value at sea level. Surface pumps are excellent for pushing water long distances in pipes and must be mounted no higher than 10 feet above your water source.

Elements needed for surface pumping includes In-line filters, to remove grit and protect your pump, foot-valve pump to prime your pump, and a Run-Dry switch to automatically shut off your pump in case it runs dry. In-line filters are usually in 10" and 30" cartridges and are placed inline between your Intake hose (submerged) and the pump.

Chapter Two - Defining Step-by-Step the Best Solar Water Pumping System for your Job

Now that we've had an overview of Solar water pumping let's take a few examples to illustrate the differences. Reading this book suggests you have a water pumping project in mind. Is your source of water from a well, or from a shallow source? The following steps will define your pumping needs and gives you the basis to choose the best hardware for the job.

Step One: Submersible or Surface Pump?

If your source water is from a well you'll use a Submersible pump. If your water source is shallow in depth, from a tank, cistern, pond, creek, stream, lake, or small river then you'll need a Surface Pump.

Step Two: What is the Height I need to pump my water, the "Rise?"

Next, let's figure out the "Rise." If you're going to pump from a well, then the rise will be the water table depth, (the depth of the water in the well) plus a margin of error, add 20 feet to your depth), or add more if you suspect the water level will drop during daily pumping. Be sure to add any additional height above the surface of the well, such as for a tank, or cistern. You'll size your pump based on the Total Lift you require.

Step Three: What is the Horizontal Distance I need, the "Run?"

The "Run" will be the total horizontal distance you want to push the water irregardless of up and downs in the land. For surface pumps, the Slow Pump options, more to come later are capable of pushing water many miles. If your water pumping project has a large horizontal "Run," then specific surface pumps are the best option.

Step Four: How much water do I need to pump and deliver Per Day?

How much water you need to pump depends on what you're doing. Are you watering a garden, or a field? Watering an orchard, or a source of water for a home, cabin, or remote site? In the example above we used watering livestock. Estimating each Head of Cattle needing 30 Gallons Per Day (GPD) we can estimate the daily herd need multiplied by the number of cattle.

Water pumps are usually rated in Gallons Per Minute (GPM). As there are 60 minutes per hour, each hour of water pumped will be 60 times GPM. If the GPM is 10 Gallons per minute, then one hour would deliver 600 Gallons. Solar electric panels, however, deliver energy over the day, and we estimate how many "Peak" hours equivalent any given location receives from the Sun. Flow rates don't give you the total solar power picture. It's vital to estimate your Total Daily needs and size your solar water pump based on Total Gallons Per Day (GPD) you require matching the energy demand of the pump, with the energy production of the solar PV panels.

Step Five: How much Solar Energy do I have on my Site?

The Sun is a powerful source of energy. Ask anyone who is stuck in the sun for a few hours. In terms of actual power, the sun is rated at Standard Test Conditions (STC). The STC condition defines the peak power density of solar energy at the surface of the Earth at 1,000 watts of power per Square Meter (about 10.5 square feet).

Note: STC also defines the amount of air-mass the sun path takes (1.5 AMO), standard temperature of 25 degrees C (77 degrees F), a wind speed of 2 meters/sec further defines a standard condition for testing, and rating solar PV panels.

To determine how much Solar Energy you have at your location look up the **Sun Peak-Hours** for your location on a Solar Map. In our examples here we're using a location in Kansas, with 5.5 Solar Peak-hours. Look up your locations solar peak-hour rating.

Raw solar energy produces, at peak condition during a clear sky, 1 Kilowatt (1,000) watts of optical power. Solar electric modules (Photovoltaic PV Panels) convert this optical energy into Direct Current (DC) with good efficiency delivering about 140 watts of electricity per square meter. Solar PV panels are "hardwired" to produce a desired voltage. Each solar "Cell" produces about 1/2 Volt DC on its own.

Amazingly, even under cloudy conditions solar cells produce good voltages. The amount of solar energy will drive the amount of "Current" the solar cells produce. More direct sun, much more current. Solar cells are interconnected to produce Solar modules which you'll use for your solar pumping project.

One square meter of sunlight is a power electrical force. Producing 140 watts, at 12 VDC, one square meter of solar energy delivers over 10 Amps of current. This is a respectable amount of power and can pump an amazing amount of water.

The Energy produced by your Solar PV array will be the Power Rating of the Panels multiplied by the Sun Peak-Hours for your location.

Once you know your Rise, Run, and Water Volume per Day desired for any given solar water pumping project now you're able to size and power this system with the appropriate solar PV system. Solar water pumping system design matches the energy demand of the pump, with the energy production of the solar array. In the chapters below we'll go over different solar water pumping systems for given depths, and water volumes.

Step Six: Select the Best Solar PV powered Water Pumping System

From the Chapters below, select the best solar PV pumping system for your project. Match the depth of your well, then select the best system example based on the Total Amount of Water you wish to deliver Each Day for that depth.

Once you know these vital statistics about your solar water pumping project your pump supplier can know how to configure your system. Your other choice is to match the systems presented in this ebook that most closely meet your water requirements. If you don't see a system powerful enough listed in this book, then go through the steps above and contact a solar pump supplier, or visit **Solardyne.com** for more information.

Chapter Three: Solar Power using Solar Photovoltaic (PV) Panels for the Power Supply

The Sun is a powerful source of energy, and ideal for the work of water pumping. Solar modules produce DC current and are well suited to extreme outdoor locations for their proven durability and reliability in the field. Solar PV panels produce strong voltages even in low light levels giving you some ability to pump even in cloudy weather, with peak outputs occurring at high sun.

The energy produced by your Solar PV panel will be the power rating multiplied by your Daily Solar peak-hour rating for your site.

Check you location with a Solar Power Map.

All voltages run "downhill." If you want to power a 12 VDC load from a solar PV panel, you'll need to produce more than 12 VDC in voltage to drive the load either from a solar panel or battery. For a 12 VDC Solar PV panel to produce a higher voltage the manufacturer will wire 36 individual solar cells in series within the module. Wiring the individual solar cells in series "Adds" the voltages producing a nominal 18 VDC. Under load, which is when you connect the pump, the voltage will drop as the solar PV panels drives the pump.

Smaller solar PV panels from 5 watts to 120 watts are usually 12 VDC Panels. If you want larger system voltages wire these panels in series. Two in series for 24 VDC. Four in series for 48 VDC. Larger solar PV panels, from 140 watts - 280 watts are wired at 24 VDC each. Wire two PV panels in series for 48 VDC systems, or four PV panels in series for 96 VDC - Ideal voltage for deeper wells.

Note: When wiring solar PV panels into arrays wire in Series to increase Voltage (current remains the same), wire in Parallel to increase Current (voltage remains the same).

Solar water pumping systems are designed to operate through a voltage range, usually 30-300 VDC. Unless otherwise specified, use 48 VDC as a system minimum. The exception to this would be when you use a specific 12, or 24 VDC solar PV small pumping system matched to a specific 12, or 24 VDC pump. The general rule is deeper depths require higher voltages.

Mounting Your Solar PV Panels - The Options

Solar panels can be mounted a variety of ways. These options include Pole mounting, Ground mounting, Roof mounting, Passive Tracking, and Active Tracking mounting.

Fixed mounts keep the solar PV panel at a specific Tilt-angle and is adjustable. To increase the output of your Solar PV array you can adjust this angle seasonally to maximize solar exposure. All Solar mounts are mounted to face South when your site is in the Northern Hemisphere, (Note: point your panels North, if you're in the Southern Hemisphere).

PV panels for water pumping need a sturdy and reliable mounting bracket. Solar PV panels can be Pole mounted, either on the Top-of-the-pole, as a masthead, or can be Side-Pole mounted. Side-Pole mounting hardware has a bracket along the bottom and top of the Solar PV panels.

Pole mounting is a great option because it keeps your panels above the ground minimizing ground effects such as increased dust. Also, wiring your panels, once they're mounted on the Mounting Hardware bracket is easier to do as crawling under the solar PV panels (J-Boxes are on the Back of the Panel) is handy.

Pole mounting your solar PV panels also makes installation easier. Smaller Solar PV panels will mount on standard 1.5" Schedule #40 pipe. Site preparation involves auguring a hole, and setting your pole in cement and aggregate.

Larger Solar PV arrays, up to 2,000 watts with Top of Pole mounting, will mount on either 2.5" Schedule #40 pipe, 3.5", or 4.5" pipe for the largest arrays. The examples below will call out the specific diameter of your mounting pipe.

For sturdiness and low cost, you can also Ground Mount your Solar panels. Ground Mounting is an A-Frame rack that allows you to Adjust your Tilt Angle.

The general ideal angle for mounting your Solar PV panels is found by taking your Latitude angle of the site, and subtract 15 degrees. Therefore, if your location has a latitude of 45 degrees, the proper tilt angle is 30 degrees as measured from horizontal.

Note: If your site is in a Tropical Location, or a very Cloudy location, the best tilt angle is no angle. Mount your panels flat. This will receive the most "Global" solar radiation, that is both direct, and indirect rays.

You can also mount your solar PV array on your roof, if your roof is near your well site. In most cases this not the case, so I'll only mention that option.

Solar energy production is increased if you're always pointing the solar PV panel toward the sun. Tracking hardware does this either in one axis - Morning through Night, or on two-axis (Altitude and Azimuth) which is most accurate.

Trackers are categorized in two types: passive, and active, respectively. Passive tracking such as with the Zomeworks gear has great robustness, and increases Solar PV panel output in energy about 25% on average. Passive-type trackers use uneven heating of internal gasses to self-adjust the panels throughout the day.

Solar water pumping loves direct sunlight. Following the sun's path, solar PV panels increase energy production - power production over time. The amount of water pumped with Solar PV panels is a direct function of energy. The more energy produced by your Solar PV array, the more water you'll pump.

Active tracking using Wattsun Active Trackers increases the output of solar PV panels as much as 35%. Using servo motors, and a solar sensor, powered by a separate solar PV array, the Wattsun trackers extract the maximum energy out of your Solar PV array. There is a cost increase for the hardware, but system performance increases dramatically. If your site is very remote, I would recommend no moving parts, and go with Top-Pole mounting requiring no maintenance potential. If you have easy access to your site, or you're in a very small foot-print, active-tracking is a great option for boosting performance.

In the sample systems listed below we'll use two Solar PV panels as examples. For smaller Solar PV panels, rated at 12 VDC each, the Dasol panels of 30, 60, 90, and 135 watt power ratings are cited. For larger Solar PV panels we'll use the REC line using the popular and widely available 250 Watt module (panel) rated at 24 VDC each.

The Solar Power systems listed below will use these solar panels, or combination of solar panels to increase voltage and/or current for more water pumped.

Chapter Four: Shallow Well Water Pumping with Solar PV from 20 to 200 Foot Depths

In this chapter we'll look at the solar power supply, and systems for pumping a shallow well of up to 200 foot depth. Smaller well pumping systems (those under 200 foot lift), such as in this example, can use the SHURFlo 9300 Submersible pump. The SHURFlo pumps are excellent for these shallow depths (up to 230') and are ideal for 12 and 24 VDC systems.

It's very easy to construct a Solar PV system to power 12 VDC, or 24 VDC systems.

Solar PV panels from 100 to 200 watts are ideal in this range and produce from 1.95 GPM for depths of 20 feet, to 1.52 GPM for depths up to 230 feet. The SHUFlo 9300 uses "positive displacement" pumps and feature a high efficiency in field conditions. The SHURFlo is a good choice for your shallow wells, but because its a "positive displacement" type of pump the diaphragms need to be replaced every 2 to 4 years depending on amount of use.

To change out the diaphragms, you'll need to turn the pump off (on the controller) to disengage the solar PV electricity to the pump. Then you'll need to pull the pump, which is to haul it up with the drop-line you've kept attached. You may need to replace the brushes, diaphragm, and valves every two years or longer, but you'll get great performance from this pump. (Note: check the connector between the cable and the pump as these sometimes corrode in harsh environments).

The SHURFlo 9300 is a submersible pump, and with the right Solar PV array can lift 1.3 GPM at 230 foot depth, and nearly 2 GPM from very shallow wells.

Small Solar PV panels for 12 and 24 VDC water pumping

For an example we'll use Dasol PV panels for the 12, and 24 VDC pumping systems. REC Solar PV panels will be used for the larger pumping systems using 250 watt solar PV panels for the examples below.

Dasol, and REC Solar PV panels are made from Monocrystalline solar cells producing the highest solar efficiencies, with strong voltage and current production over a wide range of solar conditions.

To power the SHURFlo 9300 pump you'll need to choose the right controller. There are two options: the 902-100 controller, and the 902-200 model, respectively. Each of the systems below have been selected as suggestions.

The 902-110 controller is the basic controller, and isn't waterproof so be sure to mount under cover from the elements. The controllers protect your pump from an over-current condition, as well as a low-voltage situation turning the pump off to protect the circuit. The 902-100 is ideal for 24 VDC solar PV arrays.

The 902 series controller offers a selectable switch for 12 VDC, or 24 VDC systems. This controller includes a manual on/off selector as well as inputs for three high/low water sensors and sensor wire. Sensors can hang in your well and detect a low water condition to prevent the pump from running dry which can damage your pump.

The Following is a List of Solar powered PV water pumping systems with a Parts List. Please scan down to the Well Depth, and Gallons Per Day until you find a system which closely describes your water pumping needs.

Example A:

Depth of Well 20 foot - Water delivery 1.95 Gallons Per Minute:

Parts List:
Two (2) Solar PV panels rated at 30 watts and 12 VDC each. 60 watts total array. Example PV panel: Dasol DS-A18-30, Size each: 27.2" x 13.8" x 1" Top-of-Pole Mounting Hardware for two 30 watt panels (wired in series for 24 VDC). Mounts on 1.5" Schedule #40 pipe, SHURFlo 9300 Submersible Pump. SHURFlo 902-200 Controller (Float-valves, water level sensors, optional). Drop cable, power cable (#10-2C), and foundation materials site specific

Note: To calculate water Daily output multiply GPM x 60 x peak-hours for your site. Example:(1.95 x 60 x 5.5) for Kansas at 5.5 solar peak hours as listed for that site. This comes to an average of 643 Gallons Per Day. Use your Peak-Hour rating for your site to calculate how much water this system will produce at your location.

Example B:

Depth of Well 20 foot - Water delivery 24 Gallons Per Minute:

Parts List:
Two (2) Solar PV panels rated at 250 watts and 24 VDC each, 500 watts total. Example solar PV: REC Solar PV 250PE, Size each: 65.5" x 39" x 1.5"   Top-of-Pole Mounting Hardware for two 250 watt panels (wired in series for 48 VDC). Mounts on 2.5" Schedule #40 pipe. One (1) Grundfos Submersible Pump Model 40-SQF-3 with 4" diameter rated at 24 GPM. One (1) Grundfos Controller Model: CU200 (Optional Float-switch, communications). Drop cable, power cable, and foundation materials site specific.

Daily water pumped is GPM x 60 x Peak-Hours for your site (5.5 peak hours for Kansas as example). System produces 7,920 Gallons per Day on average.

Example C:

Depth of Well 50 foot - Water delivery 27 Gallons Per Minute:

Parts List:
Four (4) Solar PV panels rated at 250 watts and 24 VDC each, 1,000 watts total. Example Solar PV panel: REC Solar PV 250PE, Size each: 65.5" x 39" x 1.5" Top-of-Pole Mounting Hardware for four 250 watt panels (wired in series for 96 VDC). Mounts on 3.5" Schedule #40 pipe. One (1) Grundfos Submersible Pump Model 40-SQF-5 with 4" diameter rated at 27 GPM. One (1)Grundfos

Controller Model: CU200 (Optional Float-switch, communications). Drop cable, power cable, and foundation materials site specific.

Daily water pumped is GPM x 60 x Peak-Hours for your site (5.5 peak hours for Kansas as example). System produces 8,910 Gallons per Day on average.

Example D:

Depth of Well 60 foot - Water deliver 1.75 Gallons Per Minute:

Parts List:
Two (2) Solar PV panels rated at 60 watts each for total of 120 watts 12 VDC each. Example PV panel: Dasol DS-A18-60, Size each: 27.2" x 26.2" x 1.38" Top-of-Pole Mounting hardware for two 60 watt panels (wired in series for 24 VDC). Mounts on 1.5" Schedule #40 pipe. One (1) SHURFlo 9300 Submersible Pump rated at 1.75 GPM. One (1) SHURFlo 902-200 Controller (float-switch, three water sensors optional). Drop Cable, Power Cable (#10-2C), and foundation materials.

Total Water delivered for our example location (Kansas) with Solar Peak-Hour rating of 5.5 Peak-Hours. Estimated daily water total is GPM x 60 x Peak-Hour rating which equals 577 Gallons per Day.

Example E:

Depth of Well 75 foot - Water delivery 8 Gallons Per Minute:

Parts List:
Two (2) Solar PV panels rated at 250 watts and 24 VDC each, 500 watts total. Example Solar PV: REC Solar PV 250PE, Size each: 65.5" x 39" x 1.5" One (1) Top-of-Pole Mounting Hardware for two 250 watt panels (wired in series for 48 VDC). Mounts on 2.5" Schedule #40 pipe. One (1) Grundfos Submersible Pump Model 11-SQF-2 with 3" diameter rated at 8 GPM. One (1)Grundfos Controller Model: CU200 (Optional Float-switch, communications). Drop cable, power cable, and foundation materials site specific

Daily water pumped is estimated 2,640 Gallons per Day.

Example F:

Depth of Well 100 feet - Water delivery 1.61 Gallons Per Minute:

Parts List:
Two (2) Solar PV panels rated at 90 watts each for a total of 180 watts at 12 VDC each. Example PV panel: Dasol DS-A18-90, Size each: 39" x 26.2" x 1.38" Top-of-Pole mounting hardware for two 90 Watt PV

panels (wired in series for 24 VDC). Mounts on 1.5"
Schedule #40 pipe. One (1) SHURFlo 9300
Submersible Pump. One (1) SHURFlo 902-200
Controller (Optional features water sensors and
float-valve). Drop Cable, Power Cable (#10-2C), and
foundation materials

Estimated daily water production 531 Gallons per
Day.

Example G:

Depth of Well 100 feet - Water delivery 6.4 Gallons
Per Minute

Parts List:
Two (2) Solar PV panels rated at 250 watts and 24
VDC each, 500 watts total. Example panel: REC
Solar PV Model: 250PE, Size each: 65.5" x 39" x 1.5"
Top-of-Pole Mounting Hardware for two 250 watt
panels (wired in series for 48 VDC). Mounts on 2.5"
Schedule #40 pipe. One (1) Grundfos Submersible
Pump Model 11-SQF-2 with 3" diameter rated at 6.4
GPM. One (1) Grundfos Controller Model: CU200
(Optional Float-switch, communications). Drop
cable, power cable, and foundation materials site
specific.

Daily water pumped is GPM x 60 x Peak-Hours for
your site (5.5 peak hours for Kansas as example).

System lifts and pumps an estimated 2,112 Gallons per Day.

Example H:

Depth of Well 100 foot - Water delivery 12 Gallons Per Minute

Parts List:
Four (4) Solar PV panels rated at 250 watts and 24 VDC each, 1,000 watts total. Example panel: REC Solar PV Model: 250PE, Size each: 65.5" x 39" x 1.5" Top-of-Pole Mounting Hardware for four 250 watt panels (wired in series for 96 VDC). Mounts on 2.5" Schedule #40 pipe. One (1) Grundfos Submersible Pump Model 11-SQF-2 with 3" diameter rated at 12 GPM. One (1)Grundfos Controller Model: CU200 (Optional Float-switch, communications). Drop cable, power cable, and foundation materials site specific.

Daily water pumped is GPM x 60 x Peak-Hours for your site (5.5 peak hours for Kansas as example). System lifts and pumps an estimated 3,960 Gallons per Day.

Example I:

Depth of Well 100 foot - Water delivery 19 Gallons Per Minute

Parts List:
Six (6) Solar PV panels rated at 250 watts and 24 VDC each, 1,500 watts total. Example solar panel: REC Solar PV Model: 250PE, Size each: 65.5" x 39" x 1.5" Top-of-Pole Mounting Hardware for Six 250 watt panels (wired in series for 144 VDC). Mounts on 3.5" Schedule #40 pipe. One (1) Grundfos Submersible Pump Model 25-SQF-7 with 3" diameter rated at 19 GPM. One (1) Grundfos Controller Model: CU200 (Optional Float-switch, communications). Drop cable, power cable, and foundation materials site specific.

Daily water pumped is GPM x 60 x Peak-Hours for your site (5.5 peak hours for Kansas as example). System lifts and pumps an estimated 6,270 Gallons per Day.

Example J:

Depth of Well 200 feet - Water delivery 1.52 Gallons Per Minute

Parts List:
Two (2) Solar PV panels rated at 135 watts each for a total of 270 watts at 12 VDC each. Example panel: Dasol DS-A18-135, Size each: 56.7" x 26.2" x 1.38" Weight: 24 lb.
Top-of-Pole mounting hardware for two 135 Watt PV panels (wired in series for 24 VDC)Mounts on 1.5"

Schedule #40 pipe. One (1) SHURFlo 9300 Submersible Pump. One (1)SHURFlo 902-200 Controller (Optional float-valve, and water sensors). Drop Cable, Power Cable (#10-2C), and foundation materials.

Water pumped per day for Kansas, with 5.5 Peak-Hours (substitute your locations peak-hour rating) equals GPM x 60 x Peak-hours. Total water pumped 500 Gallons per Day.

Example K:

Depth of Well 200 foot - Water delivery 3.8 Gallons Per Minute

Parts List:
Four (4) Solar PV panels rated at 250 watts and 24 VDC each, 1,000 watts total. Example solar panels: REC Solar PV Model: 250PE, Size each: 65.5" x 39" x 1.5" Top-of-Pole Mounting Hardware for four 250 watt panels (wired in series for 96 VDC). Mounts on 2.5" Schedule #40 pipe. One (1) Grundfos Submersible Pump Model 6-SQF-2 with 3" diameter rated at 3.8 GPM
Grundfos Controller Model: CU200 (Optional Float-switch, communications). Drop cable, power cable, and foundation materials site specific.

Daily water pumped is GPM x 60 x Peak-Hours for your site (5.5 peak hours for Kansas as example).

System lifts and pumps an estimated 1,254 Gallons per Day.

Example L:

Depth of Well 200 foot - Water delivery 7.6 Gallons Per Minute

Parts List:
Four (4) Solar PV panels rated at 250 watts and 24 VDC each, 1,000 watts total. Example solar PV: REC Solar PV Model: 250PE, Size each: 65.5" x 39" x 1.5" Top-of-Pole Mounting Hardware for four 250 watt panels (wired in series for 96 VDC). Mounts on 2.5" Schedule #40 pipe.One (1) Grundfos Submersible Pump Model 11-SQF-2 with 3" diameter rated at 7.6 GPM. One (1) Grundfos Controller Model: CU200 (Optional Float-switch, communications). Drop cable, power cable, and foundation materials site specific.

Daily water pumped is GPM x 60 x Peak-Hours for your site (5.5 peak hours for Kansas as example). System lifts and pumps an estimated 2,500 Gallons per Day.

Example M:

Depth of Well 200 foot - Water delivery 12 Gallons Per Minute

Parts List:

Six (6) Solar PV panels rated at 250 watts and 24 VDC each, 1,500 watts total. Example solar pv panel: REC Solar PV Model: 250PE, Size each: 65.5" x 39" x 1.5" Top-of-Pole Mounting Hardware for Six 250 watt panels (wired in series for 144 VDC). Mounts on 3.5" Schedule #40 pipe. One (1) Grundfos Submersible Pump Model 11-SQF-2 with 3" diameter rated at 12 GPM

Grundfos Controller Model: CU200 (Optional Float-switch, communications). Drop cable, power cable, and foundation materials site specific.

Daily water pumped is GPM x 60 x Peak-Hours for your site (5.5 peak hours for Kansas as example). System lifts and pumps an estimated 3,960 Gallons per Day.

Chapter Five - Solar Pumping Wells of 400 Foot Depth

In this chapter we'll look at solar PV powered water pumping systems for deep wells up to 400 foot depth. As we go deeper and deeper we need to increase the voltage and current produced by the Solar PV array. Water wells deeper than 200 feet require larger than 48 VDC solar arrays, and are best wired for 96 VDC. Solar PV panels are usually rated up to 600 VDC so your panels are well designed and are great at pumping water at these voltages.

Example N:

Depth of Well 400 foot - Water delivery 1.8 Gallons Per Minute

Parts List:

Two (2) Solar PV panels rated at 250 watts and 24 VDC each, 500 watts total. Example PV panels: REC Solar PV Model: 250PE, Size each: 65.5" x 39" x 1.5" Top-of-Pole Mounting Hardware for two 250 watt panels (wired in series for 48 VDC). Mounts on 2.5" Schedule #40 pipe. One (1) Grundfos Submersible Pump Model 3-SQF-3 with 3" diameter rated at 1.8 GPM. One (1) Grundfos Controller Model: CU200 (Optional Float-switch, communications). Drop cable, power cable, and foundation materials site specific.

Daily water pumped is GPM x 60 x Peak-Hours for your site (5.5 peak hours for Kansas as example). System lifts and pumps an estimated 594 Gallons per Day.

Example O:

Depth of Well 400 foot - Water delivery 4.8 Gallons Per Minute

Parts List:
Four (4) Solar PV panels rated at 250 watts and 24 VDC each, 1,000 watts total. Example panels: REC Solar PV Model: 250PE, Size each: 65.5" x 39" x 1.5" Top-of-Pole Mounting Hardware for four 250 watt panels (wired in series for 96 VDC). Mounts on 3.5" Schedule #40 pipe. One (1) Grundfos Submersible Pump Model 6-SQF-3 with 3" diameter rated at 4.8 GPM. One (1) Grundfos Controller Model: CU200

(Optional Float-switch, communications). Drop cable, power cable, and foundation materials site specific.

Daily water pumped is GPM x 60 x Peak-Hours for your site (5.5 peak hours for Kansas as example). System lifts and pumps an estimated 1,584 Gallons per Day.

Example P:

Depth of Well 400 foot - Water delivery 5.7 Gallons Per Minute

Parts List:
Six (6) Solar PV panels rated at 250 watts and 24 VDC each, 1,500 watts total. Example panels: REC Solar PV Model: 250PE, Size each: 65.5" x 39" x 1.5" Top-of-Pole Mounting Hardware for Six 250 watt panels (wired in series for 144 VDC). Mounts on 3.5" Schedule #40 pipe. One (1) Grundfos Submersible Pump Model 6-SQF-3 with 3" diameter rated at 5.7 GPM. One (1)Grundfos Controller Model: CU200 (Optional Float-switch, communications). Drop cable, power cable, and foundation materials site specific.

Daily water pumped is GPM x 60 x Peak-Hours for your site (5.5 peak hours for Kansas as example). System lifts and pumps an estimated 1,881 Gallons per Day.

Chapter Six - Solar Pumping Systems for Water Wells up to 650 Foot Depth

Listed below are several Solar PV powered water pumping systems for deep wells up to 650 foot depth. As deeper depths are pumped it may be necessary to splice your cable wire from shorter lengths. After you estimate the total length of cable you'll need for your well, (Add 20 feet for margin), try to purchase your cable in one length on a spool. However, splicing cable is sometimes required as spools can be limited to 100, or 250 foot lengths, respectively, depending on your supplier (500 foot spools do exist). Splice kits are available from your pump manufacturer, or wire supplier locally, and will be needed if your pump depth exceeds a single wire length on your spool (usually 2C with Ground wire). Splices, installed properly are sturdy, be sure to shrink wrap with a heat-gun thoroughly before using.

Example Q:

Depth of Well 650 foot - Water delivery 0.9 Gallons Per Minute

Parts List:
Two (2) Solar PV panels rated at 250 watts and 24 VDC each, 500 watts total. Example panel: REC Solar PV Model: 250PE, Size each: 65.5" x 39" x 1.5" Top-of-Pole Mounting Hardware for two 250 watt panels (wired in series for 48 VDC). Mounts on 2.5" Schedule #40 pipe. One (1)Grundfos Submersible Pump Model 3-SQF-3 with 3" diameter rated at 0.9 GPM. One (1)Grundfos Controller Model: CU200 (Optional features Float-switch, communications). Drop cable, power cable, and foundation materials site specific.

Daily water pumped is GPM x 60 x Peak-Hours for your site (5.5 peak hours for Kansas as example). System lifts and pumps an estimated 297 Gallons per Day.

Example R:

Depth of Well 650 foot - Water delivery 2.5 Gallons Per Minute

Parts List:
Four (4) Solar PV panels rated at 250 watts and 24 VDC each, 1,000 watts total. Example panels: REC Solar PV Model: 250PE, Size each: 65.5" x 39" x 1.5" Top-of-Pole Mounting Hardware for four 250 watt panels (wired in series for 96 VDC). Mounts on 3.5" Schedule #40 pipe. One (1) Grundfos Submersible Pump Model 3-SQF-3 with 3" diameter rated at 2.5 GPM. One (1) Grundfos Controller Model: CU200 (Optional Float-switch, communications). Drop cable, power cable, and foundation materials site specific.

Daily water pumped is GPM x 60 x Peak-Hours for your site (5.5 peak hours for Kansas as example). Solar Pumping System lifts and pumps an estimated 825 Gallons per Day.

Example S:

Depth of Well 650 foot - Water delivery 4.1 Gallons Per Minute

Parts List:
Six (6) Solar PV panels rated at 250 watts and 24 VDC each, 1,500 watts total. Example panels: REC Solar PV Model: 250PE, Size each: 65.5" x 39" x 1.5" Top-of-Pole Mounting Hardware for Six 250 watt panels (wired in series for 144 VDC). Mounts on 3.5" Schedule #40 pipe. One (1) Grundfos Submersible

Pump Model 6-SQF-3 with 3" diameter rated at 4.1 GPM. One (1) Grundfos Controller Model: CU200 (Optional Float-switch, communications). Drop cable, power cable, and foundation materials site specific.

Daily water pumped is GPM x 60 x Peak-Hours for your site (5.5 peak hours for Kansas as example). System lifts and pumps an estimated 1,353 Gallons per Day.

Chapter Seven - Solar Pumping Systems for Wells of 800 Foot Depth

Solar water pumping systems for depths of 800 feet require strong voltages. Solar PV panels are wired in series to "Add" voltage. To produce more current, "Amps" wire your solar panels (or sub-strings) in Parallel. The solar PV pumping systems below are configured to lift and pump the water listed in the daily Gallons per Day of water delivered. Grundfos submersible pumps are durable in the field (stainless-steel housing), and installed properly can operate 12-15 years with minimum maintenance.

If you're pumping to a tank or cistern near your Well, be sure to add the Vertical Distance you still have to pump once your water has reached the top of your well for your total lift required.

Example T:

Depth of Well 800 foot - Water delivery 1.6 Gallons Per Minute

Parts List:
Five (5) Solar PV panels rated at 250 watts and 24 VDC each, 1,250 watts total. Example solar: REC Solar PV Model: 250PE, Size each: 65.5" x 39" x 1.5" Top-of-Pole Mounting Hardware for five 250 watt panels (wired in series for 120 VDC). Mounts on 2.5" Schedule #40 pipe. One (1) Grundfos Submersible Pump Model 6-SQF-3 with 3" diameter rated at 1.6 GPM. One (1) Grundfos Controller Model: CU200 (Optional Float-switch, communications). Drop cable, power cable, and foundation materials site specific.

Daily water pumped is GPM x 60 x Peak-Hours for your site (5.5 peak hours for Kansas as example). Solar powered system lifts and pumps an estimated 528 Gallons per Day.

Example U:

Depth of Well 800 foot - Water delivery 2.5 Gallons Per Minute

Parts List:

Four (4) Solar PV panels rated at 250 watts and 24 VDC each, 1,000 watts total. Example solar panels: REC Solar PV Model: 250PE, Size each: 65.5" x 39" x 1.5" Top-of-Pole Mounting Hardware for four 250 watt panels (wired in series for 96 VDC). Mounts on 3.5" Schedule #40 pipe. One (1) Grundfos Submersible Pump Model 6-SQF-3 with 3" diameter rated at 2.5 GPM. One (1) Grundfos Controller Model: CU200 (Optional Float-switch, communications). Drop cable, power cable, and foundation materials site specific.

Daily water pumped is GPM x 60 x Peak-Hours for your site (5.5 peak hours for Kansas as example). Solar Pumping System lifts and pumps an estimated 825 Gallons per Day.

Example V:

Depth of Well 800 foot - Water delivery 3.4 Gallons Per Minute

Parts List:
Six (6) Solar PV panels rated at 250 watts and 24 VDC each, 1,500 watts total. Example solar panels: REC Solar PV Model: 250PE, Size each: 65.5" x 39" x 1.5" Top-of-Pole Mounting Hardware for Six 250 watt panels (wired in series for 144 VDC). Mounts on 3.5" Schedule #40 pipe. One (1) Grundfos Submersible Pump Model 6-SQF-3 with 3" diameter rated at 3.4 GPM. One (1) Grundfos Controller

Model: CU200 (Optional Float-switch, communications). Drop cable, power cable, and foundation materials site specific.

Daily water pumped is GPM x 60 x Peak-Hours for your site (5.5 sun peak hours for Kansas as example). Solar System lifts and pumps an estimated 1,122 Gallons per Day.

If you are looking for a Solar PV Water Pumping system with more than this capacity, and seek a larger system please visit Solardyne.com for more information regarding larger systems.

Chapter Eight - Solar Water Pumping from a Shallow Stream, Creek, Lake, Pond, River, Tank, or Cistern

In Chapters above, we looked at Submersible Pumps for well water pumping. Now let's consider pumping from a shallow source of natural water such as a creek, lake, stream, or pond, as well as pumping from tanks and cisterns.

Water quality is more of an issue with shallow sources and the basic components for your solar PV pumping system usually involve an In-Line filter, the In-take hose (the only part submerged into the water source), the pump itself, the controller to manage the system, and the solar PV panel power supply.

Unlike the typical sites for Submersible Wells which are often out in the open and offer great solar access to the solar PV panels, shallow sources of water are often under the cover of trees or bushes. If your pump is shaded, it may be necessary to site your solar PV panels a distance from the pump (the closer to the pump the better to avoid voltage drop over long distances of wire).

Surface pumps, the type used for shallow sources of water, are Not submerged, and need to be sited very Near the water source. Surface pumps remain above ground with only the intake hose submerged under water. Surface pumps require a sturdy foundation, and usually warrant a small cement pad as a foundation.

Surface water pumping is a common need. Many farms, orchards, commercial gardens, and smaller gardens use a "Gravity Feed" system for irrigation.

Remote home owners and cabins also use this approach of having a tank, or cistern that you fill with water from some source. Once filled, the farmer opens a valve near the bottom of the tank to release water for his field. In the case of remote home owners, the tank is positioned at least 40 feet (70 feet best) above the house to provide adequate pressure. The question here is the source of water to fill the tank. And, the solar PV power supply required to drive the system and deliver your water.

Solar water pumping is often used to fill tanks and cisterns from a source of water such as a creek, pond, and other source located below the tank and some distance from the home. The following shallow Surface Pumping systems and their respective solar power supplies are designed for these situations. Surface water pumping usually requires a Filter stage. Choose your filter down to 10 Micron permeability for longer pump life.

Often surface pumps require the pump to be primed before pumping can begin. If needed, most manufacturers offer a Foot-Valve pump that allows you to bring water from your source into the pump for start-up. The Foot-valve primes your pump for start-up.

Solar Water Pumping Slow and Efficient

Slow pumps take advantage of very low power needed to pump thousands of gallons per day. To achieve this high efficiency the slow pumps are milled to very high tolerances and therefore don't tolerate grit in the water.

Use In-Line filters to remove fine particulates and turbidity to protect your pump for long life. In-Line filters are rated by how fine a particulate they can filter, for Slow Pumps use 10-Micron Filters.

Water moving through a pipe encounters resistance. Pumping water too fast, at too high a rate, for a given pipe diameter increases resistance not only slowing your water delivery, but puts extra back pressure on your pump. Water pumping with a Slow Pump with 0.5" or 0.75" female outlets is designed to move the appropriate amount of water for a given lift, flow-rate, and solar power supply.

Solar powered slow pumping systems are well suited to 12, 24, and 48 VDC solar power systems. However, to drive slow pumps directly from your solar PV array you need to use the correct Controller. In start-up phase, most 12, 24, and 48 VDC solar power systems need a Linear Current Booster (LCB).

The LCB booster (included in the Controller) matches the voltage and current from your solar PV panel to the voltage and current draw from the pump. The booster also builds up sufficient charge to help in Start-Up mode where pumps always draw a sharp peak of current.

The Dankoff DSP-200 LCB Pump Controller is ideal for 12, and 24 VDC pumping systems us to 200 watts peak power. Linear current boosters (LCB) add great efficiency in low sunlight levels.

The Example solar powered systems will list the appropriate hardware for the given Lift (Rise), and linear distance through pipe (Run), and (Gallons per Day) for a given situation. Scroll down until you find a system most similar to your project.

Browse down the sample systems until you find the one closes to your water needs. These examples give you a feel for the specific pump, and the solar power supply you need to pump a given lift and distance for your project.

Example W:

Rise (Total Lift): 20 feet
Run (Total Distance through Pipe): Up to 4 miles

Shallow Water Source: Pond, Creek, Stream, Lake, small River, Tank, or Cistern - Water delivery rate 9.3 Gallons Per Minute

Parts List:
Two (2) Solar PV panel rated at 135 watts at 12 VDC each, 270 Watts total. Example solar pv panels: Dasol DS-A18-135, Size each: 56.7" x 26.2" x 1.38" Top-of-Pole Mounting Hardware for two 135 watt panels (wired in series 48 VDC). Mounts on 1.5" Schedule #40 pipe (Solar panel only). One (1) Dankoff Surface Solar Force Pump Model: 3040-48PV. One (1) Dankoff Easy Install Kit for Solar Force Piston Pumps. One (1) Dankoff 30" In-Line Filter/Foot Valve Dankoff Controller Model: PPT-48-10 includes NEMA 3R Enclosure, float-switch options allow you to have an Empty Tank Float Switch and a Full Tank Float Switch. Drop cable, power cable, and foundation materials site specific.

Quart of Food-Grade 30 wt non-toxic oil. Basic Repair Kit for 3040 modules.

Daily water pumped is GPM x 60 x Peak-Hours for your site (5.5 peak hours for Kansas as example). System lifts and pumps an estimated 3,069 Gallons per Day.

Example X:

Rise (Total Lift): 100 feet
Run (Total Distance through Pipe): Up to 4 miles

Shallow Water Source: Pond, Creek, Stream, Lake, small River, Tank, or Cistern - Water delivery rate 2.3 Gallons Per Minute

Parts List:
One (1) Solar PV panel rated at 135 watts at 12 VDC each. Example solar PV: Dasol Solar PV Module DS-A18-135, Size each: 56.7" x 26.2" x 1.38" Top-of-Pole Mounting Hardware for one 135 watt panel (12 VDC). Mounts on 1.5" Schedule #40 pipe (Solar panel only). One (1) Dankoff Surface Pump Slow Pump Model: 1303. One (1) Dankoff 30" In-Line Filter/Foot Valve
Dankoff Dry-Run Switch. One (1) Dankoff Controller Model: DSP-200 includes NEMA 3R Enclosure, float-switch option. Drop cable, power cable, and foundation materials site specific.

Daily water pumped is GPM x 60 x Peak-Hours for your site (5.5 peak hours for Kansas as example). System lifts and pumps an estimated 759 Gallons per Day.

Example Y:

Rise (Total Lift): 100 feet
Run (Total Distance through Pipe): Up to 4 miles

Shallow Water Source: Pond, Creek, Stream, Lake, small River, Tank, or Cistern - Water delivery rate 9.1 Gallons Per Minute

Parts List:
Four (4) Solar PV panel rated at 135 watts at 12 VDC each, 540 Watts total. Example panels: Dasol Solar PV panels DS-A18-135, Size each: 56.7" x 26.2" x 1.38" Top-of-Pole Mounting Hardware for four 135 watt panels (wired in series 48 VDC). Mounts on 2.5" Schedule #40 pipe (Solar panel only). One (1) Dankoff Surface Solar Force Pump Model: 3040-48PV. One (1) Dankoff Easy Install Kit for Solar Force Piston Pumps. One (1) Dankoff 30" In-Line Filter/Foot Valve. One (1) Dankoff Controller Model: PPT-48-10 includes NEMA 3R Enclosure, float-switch options allow you to have an Empty Tank Float Switch and a Full Tank Float Switch. Drop cable, power cable, and foundation materials site specific. Quart of Food-Grade 30 wt non-toxic oil. Basic Repair Kit for 3040 modules

Daily water pumped is GPM x 60 x Peak-Hours for your site (5.5 peak hours for Kansas as example). System above lifts and pumps an estimated 3,000 Gallons per Day.

Example Z:

Rise (Total Lift): 200 feet,
Run (Total Distance through Pipe): Up to 4 miles

Shallow Water Source: Pond, Creek, Stream, Lake, small River, Tank, or Cistern - Water delivery rate 2.1 Gallons Per Minute

Parts List:
Two (2) Solar PV panel rated at 135 watts at 12 VDC each, 270 Watts total. Example panels: Dasol DS-A18-135, Size each: 56.7" x 26.2" x 1.38" Weight: 24 lb. Top-of-Pole Mounting Hardware for two 135 watt panels (wired in series 24 VDC). Mounts on 1.5" Schedule #40 pipe (Solar panel only). One (1) Dankoff Surface Pump Slow Pump Model: 1303. One (1) Dankoff 30" In-Line Filter/Foot Valve Dankoff Dry-Run Switch. One (1) Dankoff Controller Model: DSP-200 includes NEMA 3R Enclosure, float-switch option. Drop cable, power cable, and foundation materials site specific.

Daily water pumped is GPM x 60 x Peak-Hours for your site (5.5 peak hours for Kansas as example).

System above lifts and pumps an estimated 693 Gallons per Day.

Example AA:

Rise (Total Lift): 200 feet
Run (Total Distance through Pipe): Up to 4 miles

Shallow Water Source: Pond, Creek, Stream, Lake, small River, Tank, or Cistern - Water delivery rate 4.8 Gallons Per Minute

Parts List:
Four (4) Solar PV panel rated at 135 watts at 12 VDC each, 540 Watts total. Example PV panels: Dasol Solar PV panels DS-A18-135, Size each: 56.7" x 26.2" x 1.38" Top-of-Pole Mounting Hardware for four 135 watt panels (wired in series 48 VDC). Mounts on 2.5" Schedule #40 pipe (Solar panel only). One (1) Dankoff Surface Solar Force Pump Model: 3040-48PV. One (1) Dankoff Easy Install Kit for Solar Force Piston Pumps, Model: EZ3000, includes Brass Manifold, Ball Valve, Check Valve, Pressure Gauge, Pressure Switch, Fittings, and Hose Bib. One (1) Dankoff Controller Model: PPT-48-10 includes NEMA 3R Enclosure, float-switch options allow you to have an Empty Tank Float Switch and a Full Tank Float Switch. One (1) Float Switch Kit. One (1) Empty Tank Shut Off, Model: 11002. One (1) Float Switch Kit Full Tank Shutoff, Model: 11023. Drop cable, power cable, and foundation materials site specific.

Quart of Food-Grade 30 wt non-toxic oil (To lubricate the motor). One (1) Basic Repair Kit for 3040 modules, Model: 3522, includes one Packing Kit, Neoprene Valve Discs, Water Box Gaskets, Valve Springs with Washers/CotterPins and Cub Leathers. Input Port diameter is 1.5 inches, with Output Port diameter of 1 inch.

Daily water pumped is GPM x 60 x Peak-Hours for your site (5.5 peak hours for Kansas as example). Solar System lifts and pumps an estimated 1,584 Gallons per Day.

Example BB:

Rise (Total Lift): 400 feet
Run (Total Distance through Pipe): Up to 4 miles

Shallow Water Source: Pond, Creek, Stream, Lake, small River, Tank, or Cistern - Water delivery rate 1.1 Gallons Per Minute

Parts List:
Three (3) Solar PV panel rated at 135 watts at 12 VDC each, 405 Watts total. Example Solar PV panels: Dasol DS-A18-135, Size each: 56.7" x 26.2" x 1.38" Top-of-Pole Mounting Hardware for three 135 watt panels (wired in series 36 VDC). Mounts on 1.5" Schedule #40 pipe (Solar panel only). One (1) Dankoff Surface Pump Slow Pump Model: 1303. One (1) Dankoff 30" In-Line Filter/Foot Valve. One (1)

Dankoff Dry-Run Switch. One (1) Dankoff Controller Model: DSP-200 includes NEMA 3R Enclosure, float-switch option. Drop cable, power cable, and foundation materials site specific.

Daily water pumped is GPM x 60 x Peak-Hours for your site (5.5 peak hours for Kansas as example). System lifts and pumps an estimated 363 Gallons per Day.

Example CC:

Dankoff Solaram Diaphragm Pumps are used for commercial and light industrial water pumping. Solar PV power supplies at 24 VDC offer remarkable performance to lifting water great heights as high as 960 feet. The Solaram Diaphragm pump is Dankoff's most powerful surface pump. These diaphragm pumps are tough and ruggedly built. Tolerant to grit and running dry, these pumps offer a tough work horse for extreme locations.

Rise (Total Lift): 400 feet
Run (Total Distance through Pipe): Up to 4 miles

Shallow Water Source: Pond, Creek, Stream, Lake, small River, Tank, or Cistern - Water delivery rate 4.4 Gallons Per Minute

Parts List:

Six (6) Solar PV panel rated at 135 watts and 12 VDC each, 810 Watts total. Example pv panels: Dasol Solar PV panels DS-A18-135, Size each: 56.7" x 26.2" x 1.38" Top-of-Pole Mounting Hardware for six 135 watt panels (wired in parallel/series 24 VDC). Mounts on 2.5" Schedule #40 pipe (Solar panels only). One (1) Dankoff Solaram Diaphragm Pump Model: 23. One (1)Dankoff Solaram 30 Amp Controller for 24 VDC Solar pumps.

One (1) Dankoff 30" In-Line Filter/Foot Valve. One (1) Dankoff float-switch options allow you to have an Empty Tank Float Switch and a Full Tank Float Switch for automatic on/off. One (1) Dankoff Float-Switch Kit. Drop cable, power cable, and foundation materials site specific, plus one Quart of Food-Grade 30 wt non-toxic lubrication oil.

Daily water pumped is GPM x 60 x Peak-Hours for your site (5.5 peak hours for Kansas as example). System lifts and pumps an estimated 1,452 Gallons per Day.

Water Storage and Pressurization

Conventional water pumping systems for Remote Homes, or Cabins, pump water from a well, or shallow water source into a "Pressure" tank that stores the water for use in the household. Pressure tanks can be mounted at ground level near the Home, or Cabin. The pressure to move the water from the tank to your Home/cabin is produced by an inflatable bladder inside the tank that pushes the water through your home pipes. This inflatable pressure is powered by the onsite solar/wind power supply, and is in addition to the solar power used in pumping the water to the tank.

Another approach, only using solar water pumping, employs Gravity to produce the house water pressure. The Solar PV power supply pumps water, using solar PV panels, from your water source (such as a creek nearby) to a tank placed at a higher elevation than your home. Minimum pressure for household use is obtained when the tank is located at least 40 feet above the house. To reach 30 PSI, considered normal water pressure in cities you should have your tank at least 70 feet above the house.

Solar water pumping systems are excellent for filling your storage tank, and equipped with a "Float-Switch" the pump can be turned off when your tank is full. Float-switches can be installed in tanks, and cistern up to 200 feet away from your pump controller.

Chapter Nine: Quick-Guide to Solar Water Pumping Examples in LIft, Flow-rate, and Gallons per Day

Listed above in each chapter are different Solar PV powered water pumping systems based on whether you're pumping from a Well, or from a Shallow source, Total Lift, Solar Pumped Flow Rates, and Daily water delivery in Gallons Per Day.

Solar PV Powered pumping systems for Deep Water Sources Such as Wells:

Solar Water Pumping System Examples in Well Depth, Flow rate in Gallons per Minute (GPM), and Total Daily Gallons in Gallons per Day (GPD)

A: 20 Foot Well, pumping 1.95 GPM, delivering 643 GPD

B: 20 Foot Well, pumping 24 GPM, delivering 7,920 GPD

C: 50 Foot Well, pumping 27 GPM, delivering 8,910 GPD

D: 60 Foot Well, pumping 1.75 GPM, delivering 577 GPD

E: 75 Foot Well, pumping 8 GPM, delivering 2,640 GPD

F: 100 Foot Well, pumping 1.61 GPM, delivering 531 GPD

G: 100 Foot Well, pumping 6.4 GPM, delivering 2,112 GPD

H: 100 Foot Well, pumping 12 GPM, delivering 3,960 GPD

I: 100 Foot Well, pumping 19 GPM, delivering 6,270 GPD

J: 200 Foot Well, pumping 1.52 GPM, delivering 500 GPD

K: 200 Foot Well, pumping 3.8 GPM, delivering 1,254 GPD

L: 200 Foot Well, pumping 7.6 GPM, delivering 2,500 GPD

M: 200 Foot Well, pumping 12 GPM, delivering 3,960 GPD

N: 400 Foot Well, pumping 1.8 GPM, delivering 594 GPD

O: 400 Foot Well, pumping 4.8 GPM, delivering 1,584 GPD

P: 400 Foot Well, pumping 5.7 GPM, delivering 1,881 GPD

Q: 650 Foot Well, pumping 0.9 GPM, delivering 297 GPD

R: 650 Foot Well, pumping 2.5 GPM, delivering 825 GPD

S: 650 Foot Well, pumping 4.1 GPM, delivering 1,353 GPD

T: 800 Foot Well, pumping 1.6 GPM, delivering 528 GPD

U: 800 Foot Well, pumping 2.5 GPM, delivering 825 GPD

V: 800 Foot Well, pumping 3.4 GPM, delivering 1,122 GPD

Shallow Source Water Pumping Systems:

Solar powered water pumping systems to pump water up to 4 miles distance with systems rated by Vertical Lift you must pump over, such as hills and obstacles, to go from your water source (creek, stream, pond, or lake) to your tank, or cistern.

W: 20 Foot Vertical Lift, pumping 9.3 GPM, delivering 3,069 GPD

X: 100 Foot Vertical Lift, pumping 2.3 GPM, delivering 759 GPD

Y: 100 Foot Vertical Lift, pumping 9.1 GPM, delivering 3,000 GPD

Z: 200 Foot Vertical Lift, pumping 2.15 GPM, delivering 709 GPD

AA: 200 Foot Vertical Lift, pumping 4.8 GPM, delivering 1,584 GPD

BB: 400 Foot Vertical Lift, pumping 1.1 GPM, delivering 363 GPD

CC: 400 Foot Vertical Lift, pumping 4.4 GPM, delivering 1,452 GPD

Solar powered water pumping systems are remarkable for their effectiveness with even a small amount of sunlight. Tap into the daily energy falling on your pump site to power the pump and deliver from hundreds to thousands of Gallons per Day.

Be sure to plan your solar PV water pumping project in terms of Site-Preparation, Equipment Design, Equipment Acquisition, Equipment Shipping, Equipment Installation, Solar Power Supply including Mounting Hardware, Controller, and all cables/piping/grounding wire.

Always use CAUTION when installing, and working with electrical devices. Solar PV panels produce respectable voltages and currents and all safety procedures should be followed. Be sure to Read your Installation Manual carefully, and follow the instructions to the letter.

Properly installed, and maintained, solar PV water pumping systems offer long life, great productivity, and ease of installation and operation. The intent of this book is to provide a resource for solar water pumping projects.

I hope you've enjoyed this book, and proves useful in planning your specific solar water pumping project. For additional information on larger systems, and other clean energy topics please visit **Solardyne.com** on the worldwide web.

Enjoy your Solar Water Pumping!

About the Author

Christopher Kinkaid

Christopher (Toby) Kinkaid, originally from Portland, Oregon is the founder of **Solardyne.com**, **SolarQuote.com**, and **AlgaeToday.com**, and has worked in clean energy technology for over three decades.

Kinkaid, is the inventor of the "**Helyx**" Vertical Axis Wind Generator, the "**Mariposa**" Non-imaging solar concentrator PV module (continuous operation at Sandia National Laboratory since 1994), the "**Solar Demultiplexer**" optical solar concentrating lens (Dr. James/Sandia National Laboratory 1991), and the inventor of the original "**Solar Power Pack**" (Mother Earth News, "Littlest Utility" June/July, 2001).

Kinkaid, has been an official lecturer and presenter on clean energy technology around the world including "APEC", Bangkok, Thailand, 2003, "Energy Solutions World", Tokyo, Japan, 2003, The International Biomass Conference (IBC), 2010,

Minneapolis, MN, and the Algal Biomass Organization (ABO) Conference, 2010, Phoenix, AZ.

Christopher (Toby) Kinkaid, has appeared in interviews on KOIN TV, KGW TV, and "Sustainable Today" produced in Oregon, and has served on the board of directors for the National Hydrogen Association, in Washington D.C., 1993, the Japan Satellite Communications Company (JCNET), Fukuoka, Japan, 1994-95, and Algaedyne Corporation, Preston, MN, 2010-2013.

Kinkaid, presently serves as CEO of Solardyne, LLC in Portland, Oregon, where he continues his work in Solar, Wind, and Biomass Technology, applications, research, and development.